FUNDAMENTALS OF ENGINEERING
GENERAL ENGINEERING
SAMPLE QUESTIONS & SOLUTIONS

FUNDAMENTALS OF ENGINEERING
GENERAL ENGINEERING
SAMPLE QUESTIONS & SOLUTIONS

Published by the
National Council of Examiners for Engineering and Surveying®
280 Seneca Creek Road, Clemson, SC 29631   800-250-3196   www.ncees.org

©2005 by the National Council of Examiners for Engineering and Surveying®
All rights reserved.

ISBN 1-932613-27-7

Printed in the United States of America

TABLE OF CONTENTS

Page

UPDATES TO EXAMINATION INFORMATION

For current exam specifications, errata for this book, a list of calculators that may be used at the exam, guidelines for requesting special accommodations, and other information about exams, visit the NCEES Web site at www.ncees.org.

INTRODUCTION

One of the functions of the National Council of Examiners for Engineering and Surveying (NCEES) is to develop examinations that are taken by candidates for licensure as professional engineers. NCEES has prepared this handbook to assist candidates who are preparing for the Fundamentals of Engineering (FE) examination in general engineering. NCEES is an organization established to assist and support the licensing boards that exist in all states and U.S. territories. NCEES provides these boards with uniform examinations that are valid measures of minimum competency related to the practice of engineering.

To develop reliable and valid examinations, NCEES employs procedures using the guidelines established in the *Standards for Educational and Psychological Testing* published by the American Psychological Association. These procedures are intended to maximize the fairness and quality of the examinations. To ensure that the procedures are followed, NCEES uses experienced testing specialists possessing the necessary expertise to guide the development of examinations using current testing techniques.

The examinations are prepared by committees composed of professional engineers from throughout the nation. These engineers supply the content expertise that is essential in developing examinations that are valid measures of minimum competency.

LICENSURE: AN IMPORTANT DECISION

One of the most important decisions you can make early in your engineering career is to place yourself on a professional course and plan to become licensed as a professional engineer (P.E.). The licensure of professional engineers is important to the public because of the significant role engineering plays in society. The profession regulates itself, through the licensing boards, by setting high standards for professional engineers. These high standards help protect the public by requiring that professional engineers demonstrate their competence to practice in a manner that will safeguard the public's safety and welfare.

The first step on the path to licensure as a P.E. is to take and pass the FE examination. If you are a student or a recent college graduate, you are well advised to take this step while coursework is still fresh in your mind. After passing the examination, your state board will designate you as an engineer intern (EI). In the past, the term "engineer-in-training" (EIT) has been used to recognize this step in your career.

To continue the licensure process, typically you must complete 4 years of progressive and verifiable experience that is acceptable to your licensing board. Some boards require that this experience be gained under the supervision of a professional engineer. Again, because of variations from state to state, you should contact your licensing board for information to ensure that you are on track to meet their requirements. Having met these requirements, you will be granted permission to take the Principles and Practice of Engineering (PE) examination. After you pass the PE examination, you may become licensed as a professional engineer and use the distinguished P.E. designation.

LICENSING REQUIREMENTS

Eligibility

The primary purpose of licensure is to protect the public by evaluating the qualifications of candidates seeking licensure. While examinations offer one means of measuring the competency levels of candidates, most licensing boards also screen candidates based on education and experience requirements. Because these requirements vary between boards, it would be wise to contact the appropriate board. Board addresses and telephone numbers may be obtained by visiting our Web site at www.ncees.org or by calling 800-250-3196.

Application Procedures and Deadlines

Application procedures for the examination and instructional information are available from individual boards. Requirements and fees vary among the boards, and applicants are responsible for contacting their board office. Sufficient time must be allotted to complete the application process and assemble required data.

DESCRIPTION OF EXAMINATIONS

Examination Schedule

The NCEES FE examination is offered to the boards in the spring and fall of each year. Dates of future administrations are published on the NCEES Web site at www.ncees.org. You should contact your board for specific locations of exam sites.

Examination Content

The purpose of the FE examination is to determine if the examinee has an adequate understanding of basic science, mathematics, engineering science, engineering economics, and discipline-specific subjects normally covered in coursework taken in the last 2 years of an engineering bachelor degree program. The examination identifies those applicants who have demonstrated an acceptable level of competence in these subjects.

The 8-hour FE examination is a no-choice examination in a multiple-choice format. The examination is administered in two 4-hour sessions. The morning session contains 120 questions, and the afternoon session contains 60 questions. Each question has four answer options. The examination specifications presented in this book give details of the subjects covered on the examinations.

Numerical questions are posed mostly in metric units, normally International System of Units (SI). However, because some subjects are typically taught in U.S. customary units (in.-lb) only, questions on the examination dealing with these subject areas are posed in U.S. customary units.

The FE examination is a closed-book examination. However, since engineers rely heavily on reference materials, you will be given a copy of the NCEES *FE Supplied-Reference Handbook* at the examination site. The *Handbook* contains formulas and data that examinees cannot reasonably be expected to commit to memory. The *Handbook* does not contain all the information required to answer every question on the examination. For example, basic theories, formulas, and definitions that examinees are expected to know have not been included. To familiarize yourself with the content of the *Handbook* before the examination, visit the NCEES Web site at www.ncees.org to view and download a free copy of the *Handbook*. You may also order, at minimal cost, a hard copy of the *Handbook* on the NCEES Web site or by calling NCEES Customer Service at 800-250-3196. You will not be allowed to take your copy of the *Handbook* into the examination; you must use the copy provided to you by the proctor in the examination room.

A sample examination is presented in this book. By illustrating the general content of the subject areas and formats, the questions should be helpful in preparing for the examination. Solutions are presented for all the questions. The solution presented may not be the only way to solve the question. The intent is to demonstrate the typical effort required to solve each question.

No representation is made or intended as to future examination questions, content, or subject matter.

Examination Preparation and Review
Examination development and review workshops are conducted at least twice annually by standing committees of the NCEES. Additionally, workshops are held as required to supplement the bank of questions available. The content and format of the questions are reviewed by the committee members for compliance with the specifications and to ensure the quality and fairness of the examination. These engineers are selected with the objective that they be representative of the profession in terms of geography, ethnic background, gender, and area of practice.

SCORING OF THE EXAMINATION

Both sessions of the FE examination are worth the same total number of points, and questions are weighted at one point for each morning question and two points for each afternoon question. Within each session, every question has equal weight. Your final score on the examination is arrived at by the summation of the numbers of points obtained in each session.

The score required to pass the examination varies between administrations of the examination. This acknowledges the fact that variations in difficulty may exist between different versions of the examination. The individual passing scores are set to reflect a minimum level of competency consistent with the purpose of licensing examinations. This procedure assures candidates that they have the same chance of passing the examination even though the difficulty of an examination may vary from one administration to another.

To accomplish the setting of fair passing scores that reflect the standard of minimum competency, NCEES conducts passing score studies on a regular basis. At these studies, a representative panel of engineers familiar with the candidate population uses a criterion-referenced procedure to set the passing score for the examination. The panel discusses the concept of minimum competence and develops a written standard that clearly articulates what skills and knowledge are required of engineers if they are to practice in a manner that protects the public health, safety, and welfare.

NCEES does not use fixed-percentage passing scores such as 70% or 75% because these fail to take into account the difficulty levels of different questions that make up the examinations. Similarly, NCEES avoids "grading on the curve" because licensure is designed to ensure that practitioners possess enough knowledge to perform professional activities in a manner that protects the public welfare. The key issue is whether an individual candidate is competent to practice and not whether the candidate is better or worse than other candidates.

The passing score can vary from one administration of the examination to another to reflect differences in difficulty levels of the examinations. However, the passing score is always based on the standard of minimum competency. To avoid confusion that might arise from fluctuations in the passing score, scores are converted to a standard scale which adopts 70 as the passing score. This technique of converting to a standard scale is commonly employed by testing specialists.

EXAMINATION PROCEDURES AND INSTRUCTIONS
Visit the NCEES Web site for current information about exam procedures and instructions.

Examination Materials
Before the morning and afternoon sessions, proctors will distribute examination booklets containing an answer sheet. You should not open the examination booklet until you are instructed to do so by the proctor. Read the instructions and information given on the front and back covers, and listen carefully to all the instructions the proctor reads.

The answer sheets for the multiple-choice questions are machine scored. For proper scoring, the answer spaces should be blackened completely. Since April 2002, NCEES has provided mechanical pencils with 0.7-mm HB lead to be used in the examination. You are not permitted to use any other writing instrument. If you decide to change an answer, you must erase the first answer completely. Incomplete erasures and stray marks may be read as intended answers. One side of the answer sheet is used to collect identification and biographical data. Proctors will guide you through the process of completing this portion of the answer sheet prior to taking the test. This process will take approximately 15 minutes.

Starting and Completing the Examination
You are not to open the examination booklet until instructed to do so by your proctor. If you complete the examination with more than 15 minutes remaining, you are free to leave after returning all examination materials to the proctor. Within 15 minutes of the end of the examination, you are required to remain until the end to avoid disruption to those still working and to permit orderly collection of all examination materials. Regardless of when you complete the examination, you are responsible for returning the numbered examination booklet assigned to you. Cooperate with the proctors collecting the examination materials. Nobody will be allowed to leave until the proctor has verified that all materials have been collected.

Calculators
Beginning with the April 2004 exam administration, the NCEES has strictly prohibited certain calculators from exam sites. Devices having a QWERTY keypad arrangement similar to a typewriter or keyboard are not permitted. Devices not permitted include but are not limited to palmtop, laptop, handheld, and desktop computers, calculators, databanks, data collectors, and organizers. The NCEES Web site (www.ncees.org) gives specific details on calculators.

Special Accommodations
The NCEES document *Guidelines for Requesting Religious and ADA Accommodations* explains the requirements for taking an NCEES exam with special testing accommodations. Candidates who wish to request special testing accommodations should refer to the NCEES Web site (www.ncees.org) under the "Exams" heading to find this document, along with frequently asked questions and forms for making the requests. To allow adequate evaluation time, NCEES must receive requests no later than 60 days prior to the exam administration.

EXAM SPECIFICATIONS
FOR THE MORNING SESSION

NATIONAL COUNCIL OF EXAMINERS FOR ENGINEERING AND SURVEYING

Fundamentals of Engineering (FE) Examination

Effective October 2005

- The FE examination is an 8-hour supplied-reference examination: 120 questions in the 4-hour morning session and 60 questions in the 4-hour afternoon session.
- The afternoon session is administered in the following seven modules—Chemical, Civil, Electrical, Environmental, Industrial, Mechanical, and Other/General engineering.
- Examinees work all questions in the morning session and all questions in the afternoon module they have chosen.

MORNING SESSION: (120 questions in 12 topic areas)

Topic Area	Approximate Percentage of Test Content
I. Mathematics	**15%**
A. Analytic geometry	
B. Integral calculus	
C. Matrix operations	
D. Roots of equations	
E. Vector analysis	
F. Differential equations	
G. Differential calculus	
II. Engineering Probability and Statistics	**7%**
A. Measures of central tendencies and dispersions (e.g., mean, mode, standard deviation)	
B. Probability distributions (e.g., discrete, continuous, normal, binomial)	
C. Conditional probabilities	
D. Estimation (e.g., point, confidence intervals) for a single mean	
E. Regression and curve fitting	
F. Expected value (weighted average) in decision-making	
G. Hypothesis testing	
III. Chemistry	**9%**
A. Nomenclature	
B. Oxidation and reduction	
C. Periodic table	
D. States of matter	
E. Acids and bases	
F. Equations (e.g., stoichiometry)	
G. Equilibrium	
H. Metals and nonmetals	

Topic Area

IV. **Computers**   7%
    A. Terminology (e.g., memory types, CPU, baud rates, Internet)
    B. Spreadsheets (e.g., addresses, interpretation, "what if," copying formulas)
    C. Structured programming (e.g., assignment statements, loops and branches, function calls)

V. **Ethics and Business Practices**   7%
    A. Code of ethics (professional and technical societies)
    B. Agreements and contracts
    C. Ethical versus legal
    D. Professional liability
    E. Public protection issues (e.g., licensing boards)

VI. **Engineering Economics**   8%
    A. Discounted cash flow (e.g., equivalence, PW, equivalent annual FW, rate of return)
    B. Cost (e.g., incremental, average, sunk, estimating)
    C. Analyses (e.g., breakeven, benefit-cost)
    D. Uncertainty (e.g., expected value and risk)

VII. **Engineering Mechanics (Statics and Dynamics)**   10%
    A. Resultants of force systems
    B. Centroid of area
    C. Concurrent force systems
    D. Equilibrium of rigid bodies
    E. Frames and trusses
    F. Area moments of inertia
    G. Linear motion (e.g., force, mass, acceleration, momentum)
    H. Angular motion (e.g., torque, inertia, acceleration, momentum)
    I. Friction
    J. Mass moments of inertia
    K. Impulse and momentum applied to:
        1. particles
        2. rigid bodies
    L. Work, energy, and power as applied to:
        1. particles
        2. rigid bodies

VIII. **Strength of Materials**   7%
    A. Shear and moment diagrams
    B. Stress types (e.g., normal, shear, bending, torsion)
    C. Stress strain caused by:
        1. axial loads
        2. bending loads
        3. torsion
        4. shear

Topic Area

    D. Deformations (e.g., axial, bending, torsion)
    E. Combined stresses
    F. Columns
    G. Indeterminant analysis
    H. Plastic versus elastic deformation

IX. Material Properties   **7%**
    A. Properties
        1. chemical
        2. electrical
        3. mechanical
        4. physical
    B. Corrosion mechanisms and control
    C. Materials
        1. engineered materials
        2. ferrous metals
        3. nonferrous metals

X. Fluid Mechanics   **7%**
    A. Flow measurement
    B. Fluid properties
    C. Fluid statics
    D. Energy, impulse, and momentum equations
    E. Pipe and other internal flow

XI. Electricity and Magnetism   **9%**
    A. Charge, energy, current, voltage, power
    B. Work done in moving a charge in an electric field
        (relationship between voltage and work)
    C. Force between charges
    D. Current and voltage laws (Kirchhoff, Ohm)
    E. Equivalent circuits (series, parallel)
    F. Capacitance and inductance
    G. Reactance and impedance, susceptance and admittance
    H. AC circuits
    I. Basic complex algebra

XII. Thermodynamics   **7%**
    A. Thermodynamic laws (e.g., 1st Law, 2nd Law)
    B. Energy, heat, and work
    C. Availability and reversibility
    D. Cycles
    E. Ideal gases
    F. Mixture of gases
    G. Phase changes
    H. Heat transfer
    I. Properties of:
        1. enthalpy
        2. entropy

Morning
Sample Questions

NOTE: THESE QUESTIONS REPRESENT ONLY HALF THE NUMBER OF QUESTIONS THAT APPEAR ON THE ACTUAL EXAMINATION.

MORNING SAMPLE QUESTIONS

1. If the functional form of a curve is known, differentiation can be used to determine all of the following **EXCEPT** the:

   (A) concavity of the curve

   (B) location of inflection points on the curve

   (C) number of inflection points on the curve

   (D) area under the curve between certain bounds

2. Which of the following is the general solution to the differential equation and boundary condition shown below?

   $$\frac{dy}{dt} + 5y = 0;\ y(0) = 1$$

   (A) e^{5t}

   (B) e^{-5t}

   (C) $e^{\sqrt{-5t}}$

   (D) $5e^{-5t}$

GO ON TO THE NEXT PAGE

3. If D is the differential operator, then the general solution to $(D + 2)^2 y = 0$ is:

(A) $C_1 e^{-4x}$

(B) $C_1 e^{-2x}$

(C) $e^{-4x}(C_1 + C_2 x)$

(D) $e^{-2x}(C_1 + C_2 x)$

4. A particle traveled in a straight line in such a way that its distance S from a given point on that line after time t was $S = 20t^3 - t^4$. The rate of change of acceleration at time $t = 2$ is:

(A) 72
(B) 144
(C) 192
(D) 208

5. Which of the following is a unit vector perpendicular to the plane determined by the vectors $\mathbf{A} = 2\mathbf{i} + 4\mathbf{j}$ and $\mathbf{B} = \mathbf{i} + \mathbf{j} - \mathbf{k}$?

    (A) $-2\mathbf{i} + \mathbf{j} - \mathbf{k}$

    (B) $\dfrac{1}{\sqrt{5}}(\mathbf{i} + 2\mathbf{j})$

    (C) $\dfrac{1}{\sqrt{6}}(-2\mathbf{i} + \mathbf{j} - \mathbf{k})$

    (D) $\dfrac{1}{\sqrt{6}}(-2\mathbf{i} - \mathbf{j} - \mathbf{k})$

6. If f' denotes the derivative of a function of $y = f(x)$, then $f'(x)$ is defined by:

    (A) $\lim\limits_{\Delta y \to 0} \dfrac{\Delta x}{\Delta y}$

    (B) $\lim\limits_{\Delta y \to 0} \dfrac{\Delta y}{\Delta x}$

    (C) $\lim\limits_{\Delta x \to 0} \dfrac{f(x + \Delta x) - f(x)}{\Delta x}$

    (D) $\lim\limits_{\Delta y \to 0} \dfrac{f(x - \Delta x) + f(x)}{\Delta y}$

GO ON TO THE NEXT PAGE

7. What is the area of the region in the first quadrant that is bounded by the line $y = 1$, the curve $x = y^{3/2}$, and the y-axis?

    (A) 2/5
    (B) 3/5
    (C) 2/3
    (D) 1

8. Three lines are defined by the three equations:

    $$x + y = 0$$
    $$x - y = 0$$
    $$2x + y = 1$$

    The three lines form a triangle with vertices at:

    (A) $(0, 0), \left(\dfrac{1}{3}, \dfrac{1}{3}\right), (1, -1)$

    (B) $(0, 0), \left(\dfrac{2}{3}, \dfrac{2}{3}\right), (-1, -1)$

    (C) $(1, 1), (1, -1), (2, 1)$

    (D) $(1, 1), (3, -3), (-2, -1)$

9. ✓ The value of the integral $\int_0^\pi 10\sin x\,dx$ is:

    (A)   −10
    (B)   0
    (C)   10
    (D)   20

10. You wish to estimate the mean M of a population from a sample of size n drawn from the population. For the sample, the mean is x and the standard deviation is s. The probable accuracy of the estimate improves with an increase in:

    (A) M
    (B) n
    (C) s
    (D) $M + s$

GO ON TO THE NEXT PAGE

11. A bag contains 100 balls numbered from 1 to 100. One ball is removed. What is the probability that the number on this ball is odd or greater than 80?

   (A)   0.2
   (B)   0.5
   (C)   0.6
   (D)   0.8

12. The standard deviation of the population of the three values 1, 4, and 7 is:

   (A) $\sqrt{3}$

   (B) $\sqrt{6}$

   (C)   4

   (D)   6

13. Suppose the lengths of telephone calls form a normal distribution with a mean length of 8.0 min and a standard deviation of 2.5 min. The probability that a telephone call selected at random will last more than 15.5 min is most nearly:

    (A)   0.0013
    (B)   0.0026
    (C)   0.2600
    (D)   0.9987

14. The volume (L) of 1 mol of H_2O at 546 K and 1.00 atm pressure is most nearly:

    (A)   11.2
    (B)   14.9
    (C)   22.4
    (D)   44.8

√ **15.** Consider the equation:

$$As_2O_3 + 3\ C \rightarrow 3\ CO + 2\ As$$

Atomic weights may be taken as 75 for arsenic, 16 for oxygen, and 12 for carbon. According to the equation above, the reaction of 1 standard gram-mole of As_2O_3 with carbon will result in the formation of:

(A) 1 gram-mole of As

(B) 28 grams of CO

(C) 150 grams of As

(D) a greater amount by weight of CO than of As

16. If 60 mL of NaOH solution neutralizes 40 mL of 0.50 M H_2SO_4, the concentration of the NaOH solution is most nearly:

(A) 0.80 M
(B) 0.67 M
(C) 0.45 M
(D) 0.33 M

GO ON TO THE NEXT PAGE

17. The atomic weights of sodium, oxygen, and hydrogen are 23, 16 and 1, respectively. To neutralize 4 grams of NaOH dissolved in 1 L of water requires 1 L of:

   (A)   0.001 normal HCl solution

   (B)   0.01 normal HCl solution

   (C)   0.1 normal HCl solution

   (D)   1.0 normal HCl solution

18. Consider the following equation:

$$K = \frac{[C]^2[D]^2}{[A]^4[B]}$$

The equation above is the formulation of the chemical equilibrium constant equation for which of the following reactions?

   (A) $C_2 + D_2 \leftrightarrow A_4 + B$

   (B) $4A + B \leftrightarrow 2C + 2D$

   (C) $4C + 2D \leftrightarrow 2A + B$

   (D) $A_4 + B \leftrightarrow C_2 + D_2$

19. The flowchart for a computer program contains the following segment:

```
VAR = 0
IF VAR < 5 THEN VAR = VAR + 2
OTHERWISE EXIT LOOP
LOOP
```

What is the value of VAR at the conclusion of this routine?

(A)   0
(B)   2
(C)   4
(D)   6

20. In a spreadsheet, the number in Cell A4 is set to 6. Then A5 is set to A4 + A4. This formula is copied into Cells A6 and A7. The number shown in Cell A7 is most nearly:

(A)   12
(B)   24
(C)   36
(D)   216

21. Consider the following program segment:

```
INPUT Z, N
S = 1
T = 1
FOR K = 1 TO N
T = T*Z/K
S = S + T
NEXT K
```

This segment calculates the sum:

(A) $S = 1 + ZT + 2\,ZT + 3\,ZT + \ldots + N\,ZT$

(B) $S = 1 + ZT + \dfrac{1}{2}ZT + \dfrac{1}{3}ZT + \ldots + \left(\dfrac{1}{N}\right)ZT$

(C) $S = 1 + \dfrac{Z}{1} + \dfrac{2Z}{2} + \dfrac{3Z}{3} + \ldots + \left(\dfrac{NZ}{N}\right)$

(D) $S = 1 + \dfrac{Z}{1!} + \dfrac{Z^2}{2!} + \dfrac{Z^3}{3!} + \ldots + \left(\dfrac{Z^N}{N!}\right)$

22. In a spreadsheet, Row 1 has the numbers 2, 4, 6, 8, 10, ... , 20 in Columns A–J, and Row 2 has the numbers 1, 3, 5, 7, 9, ... , 19 in the same columns. All other cells are zero except for Cell D3, which contains the formula: D1 + D$1*D2. This formula has been copied into cells D4 and D5. The number that appears in cell D4 is most nearly:

(A)   3
(B)   64
(C)   519
(D) 4,216

23. An engineer testifying as an expert witness in a product liability case should:

(A) answer as briefly as possible only those questions posed by the attorneys

(B) provide a complete and objective analysis within his or her area of competence

(C) provide an evaluation of the character of the defendant

(D) provide information on the professional background of the defendant

24. A professional engineer, originally licensed 30 years ago, is asked to evaluate a newly developed computerized control system for a public transportation system. The engineer may accept this project if:

(A) he or she is competent in the area of modern control systems

(B) his or her professional engineering license has not lapsed

(C) his or her original area of specialization was in transportation systems

(D) he or she has regularly attended meetings of a professional engineering society

25. You and your design group are competing for a multidisciplinary concept project. Your firm is the lead group in the design professional consortium formed to compete for the project. Your consortium has been selected as the first to enter fee negotiations with the project owner. During the negotiations, the amount you have to cut from your fee to be awarded the contract will require dropping one of the consortium members whose staff has special capabilities not available from the staff of the remaining consortium members. Can your remaining consortium ethically accept the contract?

    (A) No, because an engineer may not accept a contract to coordinate a project with other professional firms providing capabilities and services that must be provided by hired consultants.

    (B) Yes, if your remaining consortium members hire a few new lower-cost employees to do the special work that would have been provided by the consortium member that has been dropped.

    (C) No, not if the owner is left with the impression that the consortium is still fully qualified to perform all the required tasks.

    (D) Yes, if in accepting an assignment to coordinate the project, a single person will sign and seal all the documents for the entire work of the consortium.

26. You have an on-site job interview to follow up on an on-campus interview with Company A. Just before you fly to the interview, you get a call from Company B asking you to come for an on-site interview at their offices in the same city. When you inform them of your interview with Company A, they suggest you stop in after that. Company A has already paid for your airfare and, at the conclusion of your interview with them, issues you reimbursement forms for the balance of your trip expenses with instructions to file for all your trip expenses. When you inform them of your added interview stop at Company B, they tell you to go ahead and charge the entire cost of the trip to Company A. You interview with Company B, and at the conclusion, they give you travel reimbursement forms with instructions to file for all your trip expenses. When you inform them of the instructions of Company A, they tell you that the only expenses requiring receipts are airfare and hotel rooms, so you should still file for all the other expenses with them even if Company A is paying for it because students always need a little spending money. What should you do?

    (A) Try to divide the expenses between both firms as best you can.

    (B) Do as both recruiting officers told you. It is their money and their travel policies.

    (C) File for travel expenses with only one firm.

    (D) Tell all your classmates to sign up to interview with these firms for the trips.

27. A company can manufacture a product using hand tools. Costs will be $1,000 for tools and a $1.50 manufacturing cost per unit. As an alternative, an automated system will cost $15,000 with a $0.50 manufacturing cost per unit. With an anticipated annual volume of 5,000 units and neglecting interest, the breakeven point (years) is most nearly:

(A)   2.8
(B)   3.6
(C)   15.0
(D)   never

28. A printer costs $900. Its salvage value after 5 years is $300. Annual maintenance is $50. If the interest rate is 8%, the equivalent uniform annual cost is most nearly:

(A) $224
(B) $300
(C) $327
(D) $350

29. The need for a large-capacity water supply system is forecast to occur 4 years from now. At that time, the system required is estimated to cost $40,000. If an account earns 12% per year compounded annually, the amount that must be placed in the account at the end of each year in order to accumulate the necessary purchase price is most nearly:

(A) $8,000
(B) $8,370
(C) $9,000
(D) $10,000

GO ON TO THE NEXT PAGE

30. A project has the estimated cash flows shown below.

Year End	0	1	2	3	4
Cash Flow	−$1,100	−$400	+$1,000	+$1,000	+$1,000

Using an interest rate of 12% per year compounded annually, the annual worth of the project is most nearly:

(A) $450
(B) $361
(C) $320
(D) $226

31. You must choose between four pieces of comparable equipment based on the cash flows given below. All four pieces have a life of 8 years.

Parameter	Equipment			
	A	B	C	D
First cost	$25,000	$35,000	$20,000	$40,000
Annual costs	$8,000	$6,000	$9,000	$5,000
Salvage value	$2,500	$3,500	$2,000	$4,000

The discount rate is 12%. Ignore taxes. The most preferable top two projects and the difference between their present worth values are most nearly:

(A) A and C, $170
(B) B and D, $170
(C) A and C, $234
(D) B and D, $234

32. Referring to the figure below, the coefficient of static friction between the block and the inclined plane is 0.25. The block is in equilibrium.

As the inclined plane is raised, the block will begin to slide when:

(A) $\sin \phi = 1.0$

(B) $\cos \phi = 1.0$

(C) $\cos \phi = 0.25$

(D) $\tan \phi = 0.25$

33. A cylinder weighing 120 N rests between two frictionless walls as shown in the following figure.

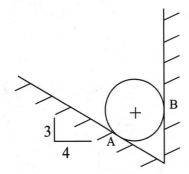

The wall reaction (N) at Point A is most nearly:

(A)   96
(B)   139
(C)   150
(D)   200

GO ON TO THE NEXT PAGE

34. Three forces act as shown below.

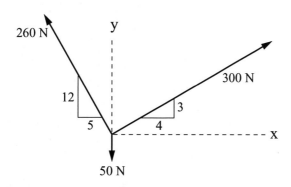

The magnitude of the resultant of the three forces (N) is most nearly:

(A) 140
(B) 191
(C) 370
(D) 396

35. In the figure below, Block A weighs 50 N, Block B weighs 80 N, and Block C weighs 100 N. The coefficient of friction at all surfaces is 0.30.

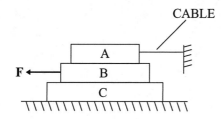

The maximum force **F** (N) that can be applied to Block B without disturbing equilibrium is most nearly:

(A) 15
(B) 54
(C) 69
(D) 84

GO ON TO THE NEXT PAGE

36.   The moment of force **F** (N•m) shown below with respect to Point p is most nearly:

(A)   31.7 ccw
(B)   31.7 cw
(C)   43.3 cw
(D)   43.3 ccw

37.   The figure below shows a simple truss.

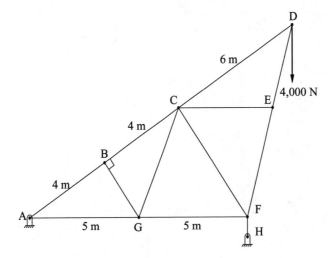

The zero-force members in the truss are:

(A)   BG, CG, CF, CE
(B)   BG, CE
(C)   CF
(D)   CG, CF

                                       GO ON TO THE NEXT PAGE

38. The beam shown below is known as a:

    (A)   cantilever beam

    (B)   statically indeterminate beam

    (C)   simply supported beam

    (D)   continuously loaded beam

39. The shear diagram for a particular beam is shown below. All lines in the diagram are straight. The bending moment at each end of the beam is zero, and there are no concentrated couples along the beam.

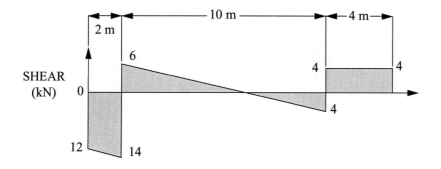

The maximum magnitude of the bending moment (kN·m) in the beam is most nearly:

    (A)   8

    (B)   16

    (C)   18

    (D)   26

40.   The piston of a steam engine is 50 cm in diameter, and the maximum steam gage pressure is 1.4 MPa. If the design stress for the piston rod is 68 MPa, its cross-sectional area (m^2) should be most nearly:

(A) 40.4×10^{-4}
(B) 98.8×10^{-4}
(C) 228.0×10^{-4}
(D) 323.0×10^{-4}

41.   A shaft of wood is to be used in a certain process. If the allowable shearing stress parallel to the grain of the wood is 840 kN/m^2, the torque (N·m) transmitted by a 200-mm-diameter shaft with the grain parallel to the neutral axis is most nearly:

(A)   500
(B)   1,200
(C)   1,320
(D)   1,500

42.   The Euler formula for columns deals with:

(A)   relatively short columns

(B)   shear stress

(C)   tensile stress

(D)   elastic buckling

GO ON TO THE NEXT PAGE

43. The mechanical deformation of a material above its recrystallization temperature is commonly known as:

    (A)   hot working

    (B)   strain aging

    (C)   grain growth

    (D)   cold working

44. In general, a metal with high hardness will also have:

    (A)   good formability

    (B)   high impact strength

    (C)   high electrical conductivity

    (D)   high yield strength

45. Glass is said to be an amorphous material. This means that it:

    (A)   has a high melting point

    (B)   is a supercooled vapor

    (C)   has large cubic crystals

    (D)   has no apparent crystal structure

46. If an aluminum crimp connector were used to connect a copper wire to a battery, what would you expect to happen?

(A) The copper wire only will corrode.

(B) The aluminum connector only will corrode.

(C) Both will corrode.

(D) Nothing

47. The rectangular homogeneous gate shown below is 3.00 m high × 1.00 m wide and has a frictionless hinge at the bottom. If the fluid on the left side of the gate has a density of 1,600 kg/m^3, the magnitude of the force **F** (kN) required to keep the gate closed is most nearly:

(A)   0
(B)   22
(C)   24
(D)   220

GO ON TO THE NEXT PAGE

48.   Which of the following statements is true of viscosity?

   (A)   It is the ratio of inertial to viscous force.

   (B)   It always has a large effect on the value of the friction factor.

   (C)   It is the ratio of the shear stress to the rate of shear deformation.

   (D)   It is usually low when turbulent forces predominate.

49.   A horizontal jet of water (density = 1,000 kg/m^3) is deflected perpendicularly to the original jet stream by a plate as shown below.

The magnitude of force **F** (kN) required to hold the plate in place is most nearly:

   (A)   4.5
   (B)   9.0
   (C)   45.0
   (D)   90.0

50. Which of the following statements about flow through an insulated valve is most accurate?

(A)   The enthalpy rises.

(B)   The upstream and downstream enthalpies are equal.

(C)   Temperature increases sharply.

(D)   Pressure increases sharply.

51. The pitot tube shown below is placed at a point where the velocity is 2.0 m/s. The specific gravity of the fluid is 2.0, and the upper portion of the manometer contains air. The reading h (m) on the manometer is most nearly:

(A)   20.0
(B)   10.0
(C)   0.40
(D)   0.20

GO ON TO THE NEXT PAGE

52.   If the complex power is 1,500 VA with a power factor of 0.866 lagging, the reactive power (VAR) is most nearly:

(A)   0
(B)   750
(C)   1,300
(D)   1,500

53.   Series-connected circuit elements are shown in the figure below.

30 Ω

j90 Ω

−j50 Ω

Which of the following impedance diagrams is correct according to conventional notation?

(A)   30 Ω

143 Ω   140 Ω

(B)   30 Ω

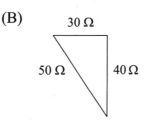

50 Ω   40 Ω

(C)

143 Ω   140 Ω

30 Ω

(D)

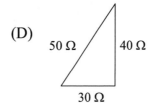

50 Ω   40 Ω

30 Ω

GO ON TO THE NEXT PAGE

54. A 10-μF capacitor has been charged to a potential of 150 V. A resistor of 25 Ω is then connected across the capacitor through a switch. When the switch is closed for ten time constants, the total energy (joules) dissipated by the resistor is most nearly:

(A) 1.0×10^{-7}
(B) 1.1×10^{-1}
(C) 9.0×10^{1}
(D) 9.0×10^{3}

55. The connecting wires and the battery in the circuit shown below have negligible resistance.

The current (amperes) through the 6-Ω resistor is most nearly:

(A) 1/3
(B) 1/2
(C) 1
(D) 3/2

56. The term $\dfrac{(1-i)^2}{(1+i)^2}$, where $i = \sqrt{-1}$, is most nearly:

(A) -1
(B) $-1 + i$
(C) 0
(D) $1 + i$

57. An insulated tank contains half liquid and half vapor by volume in equilibrium. The release of a small quantity of the vapor without the addition of heat will cause:

(A)   evaporation of some liquid in the tank

(B)   superheating of the vapor in the tank

(C)   a rise in temperature

(D)   an increase in enthalpy

58. The heat transfer during an adiabatic process is:

(A)   reversible

(B)   irreversible

(C)   dependent on temperature

(D)   zero

59.   An isentropic process is one which:

   (A)   is adiabatic but not reversible

   (B)   is reversible but not adiabatic

   (C)   is adiabatic and reversible

   (D)   occurs at constant pressure and temperature

60.   The universal gas constant is 8.314 kJ/(kmol·K). The gas constant [(kJ/(kg·K)] of a gas having a molecular weight of 44 is most nearly:

   (A)   0.19
   (B)   0.38
   (C)   0.55
   (D)   5.3

**IF YOU FINISH BEFORE TIME IS CALLED, YOU MAY WISH
TO CHECK YOUR WORK ON THIS TEST.**

MORNING SOLUTIONS

ANSWERS TO THE MORNING SAMPLE QUESTIONS

Detailed solutions for each question begin on the next page.

QUESTION	ANSWER	QUESTION	ANSWER
1	D	31	B
2	B	32	D
3	D	33	C
4	A	34	D
5	C	35	B
6	C	36	A
7	A	37	A
8	A	38	A
9	D	39	D
10	B	40	A
11	C	41	C
12	B	42	D
13	A	43	A
14	D	44	D
15	C	45	D
16	B	46	B
17	C	47	C
18	B	48	C
19	D	49	B
20	B	50	B
21	D	51	D
22	C	52	B
23	B	53	D
24	A	54	B
25	A	55	A
26	A	56	A
27	A	57	A
28	A	58	D
29	B	59	C
30	D	60	A

1. The area under a curve is determined by integration, not differentiation.

   THE CORRECT ANSWER IS: (D)

2. The characteristic equation for a first-order linear homogeneous differential equation is:

   $$r + 5 = 0$$

   which has a root at $r = -5$.

   Refer to Differential Equations in the Mathematics section of the *FE Reference Handbook*. The form of the solution is then:

   $$y = C e^{-\alpha t} \text{ where } \alpha = a \quad \text{and} \quad a = 5 \text{ for this problem}$$

   C is determined from the boundary condition.

   $$1 = C e^{-5(0)}$$
   $$C = 1$$

   Then, $y = e^{-5t}$

   THE CORRECT ANSWER IS: (B)

3. Refer to Differential Equations in the Mathematics section of the *FE Reference Handbook*. The characteristic equation for a second-order linear homogeneous differential equation is:

   $$r^2 + 2ar + b = 0$$

   In this problem, $D^2 + 4D + 4 = 0$, so:

   $$2a = 4 \text{ or } a = 2 \text{ and } b = 4$$

   In solving the characteristic equation, it is noted that there are repeated real roots: $r_1 = r_2 = -2$

   Because $a^2 = b$, the solution for this critically damped system is:

   $$y(x) = (C_1 + C_2 x) e^{-2x}$$

   THE CORRECT ANSWER IS: (D)

4. First, the velocity is:

$$V = S' = 60t^2 - 4t^3$$

Then, the acceleration is:

$$A = S'' = 120t - 12t^2$$

Finally, the rate of change of acceleration is:

$$A' = S''' = 120 - 24t$$

When $t = 2$:

$$A' = 120 - 24(2) = 120 - 48 = 72$$

THE CORRECT ANSWER IS: (A)

5. The cross product of vectors **A** and **B** is a vector perpendicular to **A** and **B**.

$$\begin{vmatrix} \mathbf{i} & \mathbf{j} & \mathbf{k} \\ 2 & 4 & 0 \\ 1 & 1 & -1 \end{vmatrix} = \mathbf{i}(-4) - \mathbf{j}(-2 - 0) + \mathbf{k}(2 - 4) = -4\mathbf{i} + 2\mathbf{j} - 2\mathbf{k}$$

To obtain a unit vector, divide by the magnitude.

$$\text{Magnitude} = \sqrt{(-4)^2 + 2^2 + (-2)^2} = \sqrt{24} = 2\sqrt{6}$$

$$\frac{-4\mathbf{i} + 2\mathbf{j} - 2\mathbf{k}}{2\sqrt{6}} = \frac{-2\mathbf{i} + \mathbf{j} - \mathbf{k}}{\sqrt{6}}$$

THE CORRECT ANSWER IS: (C)

6. Refer to Differential Calculus in the Mathematics section of the *FE Reference Handbook*.

THE CORRECT ANSWER IS: (C)

7. Define a differential strip with length $(x - 0)$ and height dy.

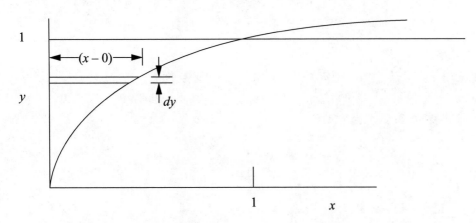

$$\int dA = \int_0^1 x\,dy = \int_0^1 y^{3/2}\,dy = \frac{y^{5/2}}{5/2}\Big|_0^1 = \frac{2}{5}$$

THE CORRECT ANSWER IS: (A)

8.

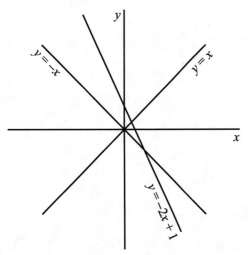

$y = -x$

$y = x$

$y = -2x + 1$

from graph one, intersection is at (0,0), so (C) and (D) are incorrect.

Also, second intersection is at (1,–1), so key has to be (A).

(0,0) (1/3, 1/3) and (1,–1)

THE CORRECT ANSWER IS: (A)

9. $$\int_{0}^{\pi} 10\sin x\,dx = 10\left[-\cos x\Big|_{0}^{\pi}\right]$$

$$= 10\left[-\cos \pi - \left(-\cos 0\right)\right]$$

$$= 10[1+1]$$

$$= 20$$

THE CORRECT ANSWER IS: (D)

10. Accuracy increases with increasing sample size.

   THE CORRECT ANSWER IS: (B)

11. The key word is **OR**. What is the probability that the number is odd **OR** greater than 80? Refer to Property 2 given under Probability and Statistics in the Mathematics section of the *FE Reference Handbook*.

   $P(A + B) = P(A) + P(B) - P(A,B)$

   Event A is removing a ball with an odd number.
   $P(A) = 50/100 = 0.5$

   Event B is removing a ball with a number greater than 80.
   $P(B) = 20/100 = 0.2$

   Event A,B is removing a ball with an odd number that is greater than 80.

   There are ten such balls.
   $P(A,B) = 10/100 = 0.1$

   Also $P(A,B) = P(A) \times P(B) = 0.5 \times 0.2 = 0.1$

   $P(A + B) = 0.5 + 0.2 - (0.5 \times 0.2) = 0.6$

   THE CORRECT ANSWER IS: (C)

12.

x	$x - \bar{x}$	$(x - \bar{x})^2$
1	−3	9
4	0	0
7	3	9
$\sum = 12$		$\sum = 18$

$$\bar{X} = \frac{12}{3} = 4$$

$$\sigma = \sqrt{\frac{18}{3}} = \sqrt{6}$$

THE CORRECT ANSWER IS: (B)

13. $8 - 15.5 = 7.5$

$$\frac{7.5}{2.5} = 3 \text{ standard deviations}$$

From the Unit Normal Distribution Table in the Mathematics section of the *FE Reference Handbook*.

For X = 3, R(X) = 0.0013

THE CORRECT ANSWER IS: (A)

14. PV = nRT

1V = (1)(0.08206)(546)

solve for V = 44.8 L

THE CORRECT ANSWER IS: (D)

15. 2 moles of As × 75 g/mole of As = 150 g of As

THE CORRECT ANSWER IS: (C)

16. $H_2SO_4 + 2\,NaOH \rightarrow Na_2SO_4 + 2\,H_2O$

0.5 M H_2SO_4 = 1.0 N H_2SO_4

1.0 M NaOH = 1.0 N NaOH

40 mL of 1.0 N H_2SO_4 = 60 mL of x N NaOH

$40 \times 1 = 60x$

x = 40/60 = 0.67 N = 0.67 M NaOH

THE CORRECT ANSWER IS: (B)

17. The molecular weight of NaOH is 40 g; therefore, 4 g/L of NaOH will form 1 L of 0.1 normal NaOH solution. One liter of 0.1 normal HCl solution is required to neutralize the NaOH.

THE CORRECT ANSWER IS: (C)

18. Refer to the Chemistry section of the *FE Reference Handbook* for the equilibrium constant of a chemical reaction.

$4A + B \leftrightarrow 2C + 2D$

THE CORRECT ANSWER IS: (B)

19.

Step	VAR
1	0
2	2
3	4
4	6
EXIT LOOP	

At the conclusion of the routine, VAR = 6.

THE CORRECT ANSWER IS: (D)

20.

Row	Column A	Value of A
4	6	6
5	A4 + A4	12
6	A5 + A4	18
7	A6 + A4	24

THE CORRECT ANSWER IS: (B)

21.

Step	Variables				
	Z	N	T	K	S
1	Z	N	.	.	.
2	Z	N	1	.	1
3	Z	N	1	1	1
.	Z	N	Z	1	1
.			Z	1	$1 + Z$

(NEXT K)

			$\dfrac{Z^2}{2}$	2	$\dfrac{1 + Z + Z^2}{2}$

(NEXT K)

			$\dfrac{Z^3}{(2)(3)}$	3	$\dfrac{1 + Z + Z^2}{2} + Z^3$ over $(2)(3)$

$$\dfrac{\dfrac{1 + Z + Z^2}{2} + Z^3}{(2)(3)}$$

(NEXT K)

			$\dfrac{Z^4}{(2)(3)(4)}$	4	

$$\dfrac{\dfrac{\dfrac{1 + Z + Z^2}{2} + Z^3}{(2)(3)} + Z^3}{(2)(3)(4)}$$

$\therefore$ The sequence is: $S = 1 + \dfrac{Z}{1!} + \dfrac{Z^2}{2!} + \dfrac{Z^3}{3!} + \dfrac{Z^4}{4!} + ... + \dfrac{Z^N}{N!}$

THE CORRECT ANSWER IS: (D)

22.

Rows	Columns				
	A	B	C	D	E
1	2	4	6	8	10
2	1	3	5	7	9
3				64	
4				519	
5					

D3: D1 + D$1 * D2 = 8 + 8(7) = 64

D4: D2 + D$1 * D3 = 7 + 8(64) = 519

THE CORRECT ANSWER IS: (C)

23. Refer to the NCEES *Model Rules of Professional Conduct*, Section A.4., in the Ethics section of the *FE Reference Handbook*.

THE CORRECT ANSWER IS: (B)

24. Refer to the NCEES *Model Rules of Professional Conduct*, Section B.1., in the Ethics section of the *FE Reference Handbook*.

THE CORRECT ANSWER IS: (A)

25. Refer to the NCEES *Model Rules of Professional Conduct*, Section B.3. and Section C.1., in the Ethics section of the *FE Reference Handbook*.

THE CORRECT ANSWER IS: (A)

26. Refer to the NCEES *Model Rules of Professional Conduct*, Section B.5. and Section B.6., in the Ethics section of the *FE Reference Handbook*.

THE CORRECT ANSWER IS: (A)

27. $1.50 (5,000) = $7,500

$0.50 (5,000) = $2,500

Annual savings = $7,500 − $2,500 = $5,000

Additional investment = $15,000 − $1,000 = $14,000

Payback = $14,000/$5,000 = 2.8 years

THE CORRECT ANSWER IS: (A)

28.   Annual cost: = $900(A/P, 8%, 5) + $50 − $300(A/F, 8%, 5)

= $900(0.2505) + $50 − $300(0.1705)

= $225.45 + $50 − $51.15

= $224.30

THE CORRECT ANSWER IS: (A)

29.

$$A = F(A/F, i, n) = 40,000(A/F, 12\%, 4) = \$8,369 \text{ per year}$$

THE CORRECT ANSWER IS: (B)

30.

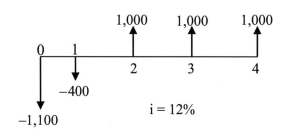

$$PW = -1,100 - 400\,(P/F, 12\%, 1) + 1,000\,(P/F, 12\%, 2)$$

$$+ 1,000\,(P/F, 12\%, 3) + 1,000\,(P/F, 12\%, 4)$$

$$= -1,100 - 400\,(0.8929) + 1,000\,(0.7972) + 1,000\,(0.7118) + 1,000\,(0.6355)$$

$$= 687.34$$

$$A = PW\,(A/P, 12\%, 4) = 687.34\,(0.3292)$$

$$= \$226 \text{ per year}$$

THE CORRECT ANSWER IS: (D)

31.   The easiest way to solve this problem is to look at the present worth of each alternative.

The present worth values are all given by
  PW = First Cost + Annual Cost × (P/A, 12%, 8) – Salvage Value × (P/F, 12%, 8)
     = First Cost + Annual Cost × 4.9676 – Salvage Value × 0.4039

Then PW(A) = $63,731
  PW(B) = $63,392
  PW(C) = $63,901
  PW(D) = $63,222

The cash flows are all costs, so the most preferable two projects, those with the lowest present worth costs, are B and D, and the difference between them is $170.

THE CORRECT ANSWER IS: (B)

32. Normal to the plane:

$$\Sigma F_n = 0: N - mg \cos \phi = 0 \rightarrow N = mg \cos \phi$$

Tangent to the plane:

$$\Sigma F_t = 0: -mg \sin \phi + \mu N = 0$$

$$\therefore -mg \sin \phi + \mu mg \cos \phi = 0$$

$$\frac{\sin \phi}{\cos \phi} = \tan \phi = \mu$$

$$\tan \phi = 0.25$$

THE CORRECT ANSWER IS: (D)

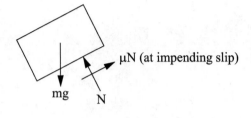

33. $\Sigma F_y = 0 = -120 + \dfrac{4}{5} A$

$A = 150 \text{ N}$

THE CORRECT ANSWER IS: (C)

34.

$$R_y = \Sigma F_y = \frac{12}{13}(260) + \frac{3}{5}(300) - 50 = 370$$

$$R_x = \Sigma F_x = -\frac{5}{13}(260) + \frac{4}{5}(300) = 140$$

$$R = \sqrt{R_x^2 + R_y^2} = \sqrt{370^2 + 140^2}$$

$$R = 396 \text{ N}$$

THE CORRECT ANSWER IS: (D)

35.

$$\Sigma F_y = 0 = -50 + N_A$$
$$N_A = 50 \text{ N}$$

$$\Sigma F_y = 0 = -50 - 80 + N_B$$
$$N_B = 130 \text{ N}$$

$$\Sigma F_y = 0 = -130 - 100 + N_C$$
$$N_C = 230 \text{ N}$$

Assume Blocks A and C remain stationary.

$$\Sigma F_x = 0 = -F' + 0.3(50) + 0.3(130)$$
$$F' = 54 \text{ N}$$

Assume Blocks B and C move.

50
0.3(50)
F''
BLOCK B
BLOCK C
0.3(230)
230

$$\Sigma F_x = 0 = -F'' + 0.3(50) + 0.3(230)$$
$$F'' = 84 \text{ N}$$

$\therefore$ **F** = 54 N where A and C remain stationary.

THE CORRECT ANSWER IS: (B)

36. $F_H = 500 \cos 30° = 433$

$F_V = 500 \sin 30° = 250$

$M_P = 250(0.30) - 433(0.10) = 31.7 \text{ N•m ccw}$

THE CORRECT ANSWER IS: (A)

37. Zero-force members usually occur at joints where members are aligned as follows:

That is, joints where two members are along the same line (OA and OC) and the third member is at some arbitrary angle. That member (OB) is a zero-force member because the forces in OA and OC must be equal and opposite.

For this specific problem, we immediately examine Joints B and E:

BG is zero-force member   CE is zero-force member

Now, examine Joint G. Since BG is zero-force member, the joint effectively looks like:

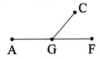

and, therefore, CG is another zero-force member.

Finally, examine Joint C. Since both CG and CE are zero force members, the joint effectively looks like:

and, therefore, CF is another zero-force member. Thus, BG, CE, CG, CF are the zero-force members.

THE CORRECT ANSWER IS: (A)

38. By definition of a cantilever beam, it is NOT statically indeterminate, it is completely supported, and it is loaded only at a specific point.

THE CORRECT ANSWER IS: (A)

39.
$$\frac{10\,m}{10\,kN} = \frac{x}{6\,kN}$$

$$x = 6\,m$$

Area 1 = 13(2) = 26 kN•m

$$\text{Area 2} = \frac{6(6)}{2} = 18\,kN\bullet m$$

Area 3 = 4(4) = 16 kN•m

SHEAR DIAGRAM

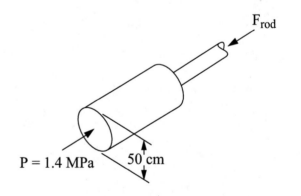

MOMENT DIAGRAM

Maximum magnitude of the bending moment is 26 kN•m.

THE CORRECT ANSWER IS: (D)

40.
$$\Sigma F = PA = \left(1.4 \times 10^6\right)\left(\frac{\pi(0.5)^2}{4}\right) = F_{rod}$$

$$F_{rod} = 275\,kN = \sigma A = 68 \times 10^6\,A$$

$$A = 40.4 \times 10^{-4}\,m^2$$

THE CORRECT ANSWER IS: (A)

58

41.

$$\tau = \frac{Tr}{J} = \frac{T\frac{d}{2}}{\frac{\pi d^4}{32}} = \frac{16T}{\pi d^3}$$

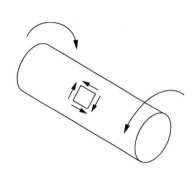

$$T = \frac{\pi d^3 \tau}{16} = \frac{\pi (0.2)^3 (840 \times 10^3)}{16}$$

$$T = 1,319 \ N \cdot m$$

THE CORRECT ANSWER IS: (C)

42. The Euler formula is used for elastic stability of relatively long columns, subjected to concentric axial loads in compression.

THE CORRECT ANSWER IS: (D)

43. The question statement is the definition of hot working.

THE CORRECT ANSWER IS: (A)

44. By definition, a metal with high hardness has a high tensile strength and a high yield strength.

THE CORRECT ANSWER IS: (D)

45. By definition, amorphous materials do not have a crystal structure.

THE CORRECT ANSWER IS: (D)

46. Aluminum is anodic relative to copper and, therefore, will corrode to protect the copper.

THE CORRECT ANSWER IS: (B)

47. The mean pressure of the fluid acting on the gate is evaluated at the mean height, and the center of pressure is 2/3 of the height from the top; thus, the total force of the fluid is:

$$F_f = \rho g \frac{H}{2}(H) = 1,600(9.807)\frac{3}{2}(3) = 70,610 \text{ N}$$

and its point of application is 1.00 m above the hinge. A moment balance about the hinge gives:

$$F(3) - F_f(1) = 0$$

$$F = \frac{F_f}{3} = \frac{70,610}{3} = 23,537 \text{ N}$$

THE CORRECT ANSWER IS: (C)

48. Refer to the Fluid Mechanics section of the *FE Reference Handbook*.

$$\tau_t = \mu\left(\frac{dv}{dy}\right)$$

where τ_t = shear stress and

$\dfrac{dv}{dy}$ = rate of shear deformation

Hence, μ is the ratio of shear stress to the rate of shear deformation.

THE CORRECT ANSWER IS: (C)

49. $Q = A_1 V_1 = (0.01 \text{ m}^2)(30 \text{ m/s})$

$= 0.3 \text{ m}^3/\text{s}$

Since the water jet is deflected perpendicularly, the force F must deflect the total horizontal momentum of the water.

$F = \rho Q V = (1,000 \text{ kg/m}^3)(0.3 \text{ m}^3/\text{s})(30 \text{ m/s}) = 9,000 \text{ N} = 9.0 \text{ kN}$

THE CORRECT ANSWER IS: (B)

50. Flow through an insulated valve closely follows a throttling process. A throttling process is at constant enthalpy.

THE CORRECT ANSWER IS: (B)

51. $\dfrac{\rho v^2}{2} = gh\left(\rho - \rho_{air}\right)$

$\therefore h = \dfrac{\rho v^2}{2g\left(\rho - \rho_{air}\right)} \approx \dfrac{v^2}{2g} \approx \dfrac{(2)^2}{(2)(9.8)} \approx 0.204 \text{ m}$

THE CORRECT ANSWER IS: (D)

52. S = apparent power
P = real power
Q = reactive power

$S = P + jQ = |S| \cos \theta + j |S| \sin \theta$

$\cos \theta = pf = 0.866$

$Q = (1,500 \text{ VA}) \sin[\cos^{-1} 0.866] = 750 \text{ VAR}$

THE CORRECT ANSWER IS: (B)

53. $Z = 30 + j90 - j50 = 30 + j40 \ \Omega$

50 Ω   40 Ω

30 Ω

THE CORRECT ANSWER IS: (D)

54. Initially, $V_C(t) = 150 \ V$

$$W_C(t) = \frac{1}{2}cV_C^2 = \frac{1}{2}(10 \times 10^{-6} F)(150 \ V)^2$$

$W_C = 0.113 \ J$ initial stored energy.

After ten time constants, all energy will be dissipated.

THE CORRECT ANSWER IS: (B)

55. $R_T = 4\Omega + 3\Omega \parallel 6\Omega = 4\Omega + 2\Omega$

$$R_T = 6\Omega \Rightarrow I_T = \frac{6 \ V}{6\Omega} = 1 \ A$$

$$I_x = \frac{3}{9}(I_T) = \frac{1}{3} A$$

6 V   4 Ω

3 Ω   6 Ω

THE CORRECT ANSWER IS: (A)

56. $\dfrac{(1-i)^2}{(1+i)^2} = \dfrac{1-2i+i^2}{1+2i+i^2} = \dfrac{1-1-2i}{1-1+2i} = \dfrac{-i}{i} = -1$

THE CORRECT ANSWER IS: (A)

57. As vapor escapes, the mass within the tank is reduced. With constant volume, the specific volume within the tank must increase. This can happen only if liquid evaporates.

THE CORRECT ANSWER IS: (A)

58. By definition, an adiabatic process is a process in which no heat is transferred.

THE CORRECT ANSWER IS: (D)

59. An isentropic process is one for which the entropy remains constant. Entropy is defined by the equation:

$$dS = \left(\frac{\delta Q}{T} \right)_{reversible}$$

The entropy will be constant if $\delta Q = 0$ and the process is reversible. It is theoretically possible for a nonadiabatic, irreversible process to have a constant entropy, but this is not one of the responses. Option (D) describes a state, not a process.

THE CORRECT ANSWER IS: (C)

60. $R = \dfrac{R_u}{M} = \dfrac{8.314}{44} = 0.1890 \dfrac{kJ}{kg \cdot K}$

THE CORRECT ANSWER IS: (A)

EXAM SPECIFICATIONS
FOR THE AFTERNOON SESSION

AFTERNOON SESSION IN OTHER/GENERAL ENGINEERING: (60 questions in 9 topic areas)

Topic Area	Approximate Percentage of Test Content

I. Advanced Engineering Mathematics   10%
A. Differential equations
B. Partial differential calculus
C. Numerical solutions (e.g., differential equations, algebraic equations)
D. Linear algebra
E. Vector analysis

II. Engineering Probability and Statistics   9%
A. Sample distributions and sizes
B. Design of experiments
C. Hypothesis testing
D. Goodness of fit (coefficient of correlation, chi square)
E. Estimation (e.g., point, confidence intervals) for two means

III. Biology   5%
A. Cellular biology (e.g., structure, growth, cell organization)
B. Toxicology (e.g., human, environmental)
C. Industrial hygiene [e.g., personnel protection equipment (PPE), carcinogens]
D. Bioprocessing (e.g., fermentation, waste treatment, digestion)

IV. Engineering Economics   10%
A. Cost estimating
B. Project selection
C. Lease/buy/make
D. Replacement analysis (e.g., optimal economic life)

V. Application of Engineering Mechanics   13%
A. Stability analysis of beams, trusses, and frames
B. Deflection analysis
C. Failure theory (e.g., static and dynamic)
D. Failure analysis (e.g., creep, fatigue, fracture, buckling)

VI. Engineering of Materials   11%
A. Material properties of:
    1. metals
    2. plastics
    3. composites
    4. concrete

	Topic Area	**Approximate Percentage of Test Content**

VII. Fluids   15%
   A. Basic hydraulics (e.g., Manning equation, Bernoulli theorem, open-channel flow, pipe flow)
   B. Laminar and turbulent flow
   C. Friction losses (e.g., pipes, valves, fittings)
   D. Flow measurement
   E. Dimensionless numbers (e.g., Reynolds number)
   F. Fluid transport systems (e.g., pipes, ducts, series/parallel operations)
   G. Pumps, turbines, and compressors
   H. Lift/drag

VIII. Electricity and Magnetism   12%
   A. Equivalent circuits (Norton, Thevenin)
   B. AC circuits (frequency domain)
   C. Network analysis (Kirchhoff laws)
   D. RLC circuits
   E. Sensors and instrumentation
   F. Electrical machines

IX. Thermodynamics and Heat Transfer   15%
   A. Thermodynamic properties (e.g., entropy, enthalpy, heat capacity)
   B. Thermodynamic processes (e.g., isothermal, adiabatic, reversible, irreversible)
   C. Equations of state (ideal and real gases)
   D. Conduction, convection, and radiation heat transfer
   E. Mass and energy balances
   F. Property and phase diagrams (e.g., T-s, h-P)
   G. Tables of thermodynamic properties
   H. Cyclic processes and efficiency (e.g., refrigeration, power)
   I. Phase equilibrium and phase change
   J. Thermodynamic equilibrium
   K. Combustion and combustion products (e.g., CO, CO_2, NOX, ash, particulates)
   L. Psychrometrics (e.g., humidity)

GENERAL
AFTERNOON SAMPLE QUESTIONS

NOTE: THESE QUESTIONS REPRESENT ONLY HALF THE NUMBER OF QUESTIONS THAT APPEAR ON THE ACTUAL EXAMINATION.

GENERAL SAMPLE QUESTIONS

1. Under certain conditions, the motion of an oscillating spring and mass is described by the differential equation:

$$\frac{d^2x}{dt^2} + 16x = 0$$

where x is the displacement of the end of the spring in meters and t is the time in seconds. At $t = 0$ s, the displacement is 0.08 m and the velocity is 0 m/s; that is:

$$x(0) = 0.08 \text{ and } x'(0) = 0$$

The solution that fits the initial conditions is:

(A) $x = -\sin 4t$

(B) $x = 0.02 \sin 4t + 0.08 \cos 4t$

(C) $x = 4 \cos 4t$

(D) $x = 0.08 \cos 4t$

2. Consider the three vectors:

$\mathbf{A} = 5\mathbf{i} - 3\mathbf{j} + 6\mathbf{k}$
$\mathbf{B} = -\mathbf{i} + 2\mathbf{j} + 2\mathbf{k}$
$\mathbf{C} = 2\mathbf{i} + 3\mathbf{j} - 4\mathbf{k}$

The product $(\mathbf{B} \times \mathbf{C}) \bullet \mathbf{A}$ is most nearly:

(A)   −112
(B)   0
(C)   28
(D)   112

3.   For the equation $f(x, y, z) = \dfrac{3}{x^2} + ye^z + z\tan z$, $\dfrac{\partial f}{\partial z}$ is most nearly:

(A) $-\dfrac{6}{x^3}$

(B) $ye^z + z\sec^2 z + \tan z$

(C) e^z

(D) $ye^z + \tan z$

4.   A set consists of four points that have the following coordinates:

x	y
2.3	5.6
1.2	4.1
4.2	8.6
6.3	10.1

Using linear least squares regression, the equation that best fits this data is:

(A) $y = 2.9 + 1.2x$

(B) $y = 2.7 + 1.2x$

(C) $y = 2.9 - 1.2x$

(D) $y = 2.5 + 1.4x$

5. A sample of resistor values indicates an equal probability distribution as follows:

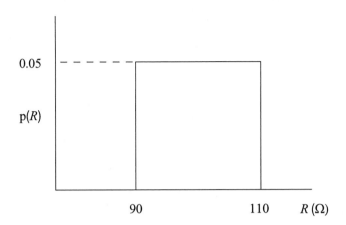

The probability that R is greater than 105 is most nearly:

(A) 20%
(B) 25%
(C) 30%
(D) 35%

6. The chemical composition of algae has been written as $C_{106}H_{263}O_{110}N_{16}P$. Atomic weights are as follows: C = 12, H = 1, N = 14, O = 16, P = 31. According to this formula, the percent by weight of nitrogen in algae is most nearly:

(A) 1.6%
(B) 6.3%
(C) 7.4%
(D) 12.6%

7.   A contractor is considering trading in an older truck crane on a new model. The data are as follows.

Contractor's ordinary income-tax bracket	40%
Purchase price of new truck crane	$150,000
Estimated life	60 mo
Initial monthly operating and maintenance (O&M)	$4,500
Monthly gradient increase in O&M	$50

Use $i = 1.5\%$ per month, compounded monthly.

The present worth after taxes of only the monthly gradient increase in O&M is most nearly:

(A) $19,760
(B) $24,020
(C) $29,650
(D) $49,400

Questions 8–9: A company is considering purchasing either Machine I or Machine II. The following data apply:

Parameter	Machine I	Machine II
First cost	$80,000	$100,000
Estimated life (years)	20	25
Salvage value	$20,000	$25,000
Other costs, including taxes, insurance, and operation (per year)	$18,000	$15,000 (first 15 years) $20,000 (next 10 years)

The annual interest rate is 10%, and all cash flows may be treated as end-of-year cash flows. Assume that the equivalent annual cost is the value of the level annuity equal to the total cost of a project.

8. If Machine II is purchased, the equivalent annual cost is most nearly:

   (A) $21,000
   (B) $21,500
   (C) $26,575
   (D) $27,080

9. The purchase of Machine I assumes an income of $25,000 per year to the company. The annual rate of return is most nearly:

   (A) 6.9%
   (B) 10.0%
   (C) 11.7%
   (D) 12.2%

10. A company plans to buy a new car for each member of its sales force. Because of the volume discount, each car will cost $21,000. The company's accountants rule that the car must be depreciated over 5 years. If the modified ACRS depreciation is used, the value of the car at the end of the fifth year will be most nearly:

(A) 0
(B) $1,218
(C) $2,415
(D) $3,633

11. A simply supported beam carries a load of 10 kN/m on a 5.0-m span. Neglecting the weight of the beam, the maximum deflection (mm) of the beam is most nearly:

(A) 10
(B) 12
(C) 15
(D) 19

$E = 200 \text{ GPa}$
$A = 7,500 \text{ mm}^2$
$I = 21.58 \times 10^6 \text{ mm}^4$

12. The maximum shear stress (MPa) in the stepped beam shown below is most nearly:

(A)   4.17
(B)   6.25
(C)   9.38
(D)   12.50

BEAM WIDTH = 12 cm

13. The critical force **P** that can be applied to the round steel bar shown in the figure below is most nearly:

(A)   40 kN
(B)   80 kN
(C)   255 kN
(D)   10 MN

GENERAL SAMPLE QUESTIONS

Questions 14–15: Refer to the equation, graph, and table below.

$$\text{Carbon equivalent} = \%\,\text{Carbon} + \frac{\%\,\text{Manganese} + \%\,\text{Chromium}}{6} + \frac{\%\,\text{Silicon} + \%\,\text{Nickel} + \%\,\text{Molybdenum}}{10}$$

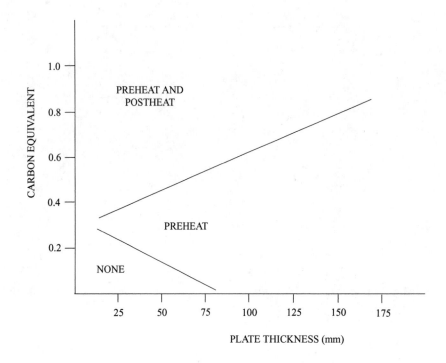

Steel Grade	Composition (%)					
	Carbon	**Manganese**	**Silicon**	**Chromium**	**Nickel**	**Molybdenum**
A36	0.25	0.85	0.28	0.05	0.03	0.07
A517	0.09	0.35	0.01	0.02	0.03	0.01
A572	0.23	1.65	0.30	0.01	0.08	0.03
A715	0.05	1.65	0.90	0.80	0.07	0.06

GO ON TO THE NEXT PAGE

14. The steel has the following chemistry: 0.25% carbon, 1.27% manganese, 0.28% silicon, 0.02% chromium, 0.05% nickel, and 0.02% molybdenum. What type of thermal treatment is necessary for weldment of a 37-mm-thick structural steel plate?

    (A)   None

    (B)   Preheat only

    (C)   Preheat and postheat

    (D)   Postheat only

15. Facilities are not available for preheating or postheating the structure being fabricated. Therefore, it is necessary to choose a grade of steel that does not require preheating or postheating. According to the table provided on the opposite page, which steel satisfies the **no-heat** treatment specification?

    (A)   A36
    (B)   A517
    (C)   A572
    (D)   A715

16. Copper and chromium are essentially insoluble in one another at room temperature. This has led to a system being investigated as a metal/metal composite. Such a composite is made with a composition of copper –30 wt% chromium. The following data apply:

Property	Copper	Chromium
Atomic number	29	24
Atomic weight (g/mol)	63.55	52.00
Electrical resistivity ($\mu\Omega\cdot$cm)	1.70	12.9
Tensile modulus (psi)	16×10^6	36×10^6
Density (g/cm^3)	8.93	7.19

The best estimate of the electrical resistivity ($\mu\Omega\cdot$cm) of the composite, ρ_c, is most nearly:

(A) 1.7
(B) 5.6
(C) 10.0
(D) 12.9

17. Air flows isothermally in a duct of constant cross-sectional area under steady-flow conditions. At one point in the duct, the pressure is 600 kPa and the velocity is 60 m/s. If the pressure is 300 kPa at another point, the velocity (m/s) is most nearly:

(A) 30
(B) 60
(C) 120
(D) 240

GO ON TO THE NEXT PAGE

Questions 18–19: The level in a retention basin is normally controlled with a pipe as shown in the figure below. The pipe has an I.D. of 30 cm. The equivalent length of the pipe (including the elbows, entrance effect, and discharge) is 6.0 m. Relative roughness is 0.0005. The fluid has the following properties:

$$\rho = 998 \text{ kg/m}^3$$
$$\mu = 0.00100 \text{ kg/(m·s)}$$

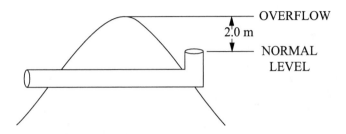

18.   Assuming a flow of 40 m³/min, the velocity (m/s) through the pipe is most nearly:

(A)   9.4
(B)   2.4
(C)   1.4
(D)   0.047

19.   Assuming a Reynolds number of 200,000, the Moody friction factor *f* is most nearly:

(A)   0.017
(B)   0.019
(C)   0.022
(D)   0.032

Questions 20–21: The figure below represents a water-flow system in which water is pumped from the lake to the storage tank and also flows from the lake through the turbine. Darcy friction factors are given for the pipe flows:

$$f = h_f \frac{D}{L} \frac{2g}{V^2} \text{ (Fanning friction factors are one-fourth as large.)}$$

Specific weight of water $9{,}800 \text{ N/m}^3$
Atmospheric pressure $1.01 \times 10^5 \text{ N/m}^2$

NOT TO SCALE

20.   With Tank A and Pipe BC filled with water to the level shown, but with no flow, the gage pressure (kPa) at Point C is most nearly:

     (A)   9
     (B)   29
     (C)   88
     (D)   190

21.   The pump is 80% efficient and is to pump water to the tank at the rate of $0.5 \text{ m}^3/\text{min}$ at a time when the tank is filled to the level shown. The pipe system from B to E is equivalent to 100 m of 75-mm-diameter pipe with a Darcy friction factor of 0.02. The power (W) delivered to the pump is most nearly:

     (A)   490
     (B)   1,250
     (C)   1,560
     (D)   93,900

GO ON TO THE NEXT PAGE

22. In the circuit shown below, the switch has been open a long time so that the current I is 10 mA.

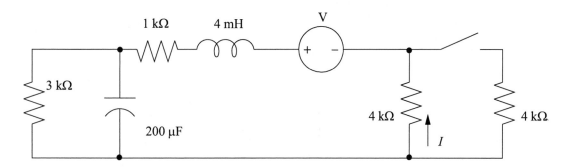

Immediately after the switch is closed, the current I (mA) is most nearly:

(A) 0
(B) 5
(C) 10
(D) 20

23. The Thevenin equivalent voltage (V) at Terminals ab in the circuit shown below is most nearly:

(A) 10
(B) 35
(C) 61
(D) 100

24. For the circuit shown below, the voltage V_2 (V) is most nearly:

(A)   2.5
(B)   4
(C)   5
(D)   10

25. A multitube heat exchanger for cooling a process fluid with water contains 25-mm O.D. tubes with a wall thickness of 3 mm. After a period of operation, the following quantities are known:

Overall heat-transfer coefficient (outside area)	700 W/(m^2·K)
Film coefficient on the shell side	1,500 W/(m^2·K)
Film coefficient on the tube side	2,500 W/(m^2·K)

Neglecting wall resistance, the combined fouling factor (m^2·K/W) based on outside area under these conditions is most nearly:

(A)   0.00015
(B)   0.00024
(C)   0.00036
(D)   0.00046

Questions 26–28: The figure below shows a Rankine cycle with water as the working medium. Disregard pressure losses in the piping, steam boiler, superheater, and kinetic and potential energy effects. Assume that there is a steady flow with adiabatic expansion in the turbine.

State	h (kJ/kg)
1	2,960.70
2	2,237.40
3	251.13
4	255.18
5	2,715.70

26. The work (kJ/kg) done by the steam on the turbine is most nearly:

    (A)   245
    (B)   723
    (C)   968
    (D)   1,080

27. The required ratio of the mass of cooling water to the mass of steam is most nearly:

    (A)   12.2
    (B)   31.9
    (C)   43.1
    (D)   181.0

28. The heat (MJ/kg) that must be added in the steam boiler is most nearly:

    (A)   2.46
    (B)   2.58
    (C)   2.72
    (D)   2.77

GO ON TO THE NEXT PAGE

Questions 29–30 are based on the Carnot power cycle shown below:

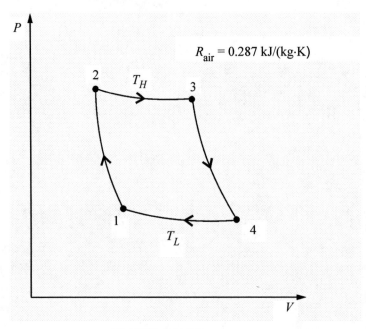

$$R_{air} = 0.287 \text{ kJ/(kg·K)}$$

NOT TO SCALE

The working medium is 1 kg of air as an ideal gas. The cycle has a thermal efficiency of 48%. The heat transfer to air during the isothermal expansion is 60 kJ. At the beginning of the isothermal expansion, the pressure is 9 bars and the volume is 0.3 m^3.

29.   The volume (m^3) at the end of the isothermal expansion process is most nearly:

    (A)   0.375
    (B)   0.451
    (C)   0.521
    (D)   0.623

30.   The heat rejected at T_L (kJ) is most nearly:

    (A)   0
    (B)   28.8
    (C)   31.2
    (D)   60

**IF YOU FINISH BEFORE TIME IS CALLED, YOU MAY
WISH TO CHECK YOUR WORK ON THIS TEST.**

GENERAL
AFTERNOON SOLUTIONS

ANSWERS TO THE GENERAL AFTERNOON SAMPLE QUESTIONS

Detailed solutions for each question begin on the next page.

QUESTION	ANSWER
1	D
2	A
3	B
4	A
5	B
6	B
7	C
8	C
9	A
10	B
11	D
12	D
13	A
14	C
15	B
16	B
17	C
18	A
19	B
20	C
21	C
22	B
23	B
24	C
25	B
26	B
27	C
28	A
29	A
30	C

1. Refer to Differential Equations in the Mathematics section of the *FE Reference Handbook*. Specifically, this is a second-order linear homogeneous differential equation.

Putting the equation into the form given in the *Handbook* yields:

$2a = 0$, so $a = 0$ and $b = 16$

From this it is seen that $a^2 < b$, so the solution is of the form (underdamped):

$$y = e^{\alpha x}(c_1 \cos \beta x + c_2 \sin \beta x)$$

where

$$\alpha = -a = 0 \quad \text{and} \quad \beta = \sqrt{b^2 - a} = \sqrt{16} = 4$$

so

$$x \quad = e^{0x}(c_1 \cos 4t + c_2 \sin 4t)$$
$$= c_1 \cos 4t + c_2 \sin 4t$$

and

$$x' = -4c_1 \sin 4t + 4c_2 \cos 4t$$

At initial condition of $x' = 0$ at $t = 0$

$$-4c_1 \sin 0 + 4c_2 \cos 0 = 0$$

$$4c_2 \quad = 0$$

$$c_2 \quad = 0$$

Therefore

$$x = c_1 \cos 4t$$

At initial condition of $x = 0.08$ at $t = 0$

$$c_1 \cos 0 = 0.08$$

$$c_1 = 0.08$$

Consequently

$$x = 0.08 \cos 4t$$

THE CORRECT ANSWER IS: (D)

2.

$$\left(\overline{\mathbf{B}} \times \overline{\mathbf{C}}\right) = \begin{vmatrix} \vec{\mathbf{i}} & \vec{\mathbf{j}} & \vec{\mathbf{k}} \\ -1 & 2 & 2 \\ 2 & 3 & -4 \end{vmatrix}$$

$$= \mathbf{i}(-8-6) + \mathbf{j}(4-4) + \mathbf{k}(-3-4)$$

$$= -14\mathbf{i} - 7\mathbf{k}$$

$$\left(\overline{\mathbf{B}} \times \overline{\mathbf{C}}\right) \cdot \mathbf{A} = \left(-14\mathbf{i} + 0\mathbf{j} - 7\mathbf{k}\right) \cdot \left(5\mathbf{i} - 3\mathbf{j} + 6\mathbf{k}\right)$$

$$= (-14)(5) + (-7)(6)$$

$$= -70 - 42 = -112$$

THE CORRECT ANSWER IS: (A)

3.

$$f(x, y, z) = \frac{3}{x^2} + y\mathrm{e}^z + z \tan z$$

$$\frac{\partial f}{\partial z} \Rightarrow \text{treat } x \text{ and } y \text{ as constants}$$

Also, refer to the table of Derivatives and Indefinite Integrals in the *FE Reference Handbook*, if necessary.

$$\frac{\partial f}{\partial z} = 0 + y\left(\mathrm{e}^z\right) + z\left(\sec^2 z\right) + \tan z (1)$$

$$= y\mathrm{e}^z + z \sec^2 z + \tan z$$

THE CORRECT ANSWER IS: (B)

4. Use equations from the Engineering Probability and Statistics section of the *FE Reference Handbook.*

$$S_{xy} = \sum x_i y_i - \left(\frac{1}{n}\right)\left(\sum x_i\right)\left(\sum y_i\right) \quad = 117.55 - \left(\frac{1}{4}\right)(14)(28.4)$$

$$= 18.15$$

$$S_{xx} = \sum x_i^2 - \left(\frac{1}{n}\right)\left(\sum x_i\right)^2 \quad = 64.06 - \left(\frac{1}{4}\right)(14)^2$$

$$= 15.06$$

$b = S_{xy}/S_{xx} = 1.2052$
$a = \bar{y} - b\bar{x} = 7.1 - 1.2052(3.5) = 2.8818$
$y = 2.88 + 1.20x$

THE CORRECT ANSWER IS: (A)

5. $$\text{PROB}(R \geq 105) = 1 - \text{F}(R = 105) = 1 - \int_{90}^{105} p(R)\,dR$$

$$= 1 - \int_{90}^{105} 0.05\,dR$$

$$= 1 - 0.05\,R\Big|_{90}^{105} = 0.25$$

$$\therefore 25\%$$

Or, area under curve:

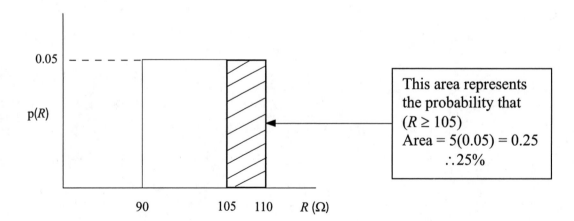

THE CORRECT ANSWER IS: (B)

6.

$16 \text{ N} \quad = 16 \times 14 \quad = 224$

$106 \text{ C} = 106 \times 12 = 1{,}272$

$263 \text{ H} = 263 \times 1 \quad = 263$

$110 \text{ O} = 16 \times 110 = 1{,}760$

$1 \text{ P} \quad = 1 \times 31 \quad = 31$

$224/3{,}550 = 6.31\%$

THE CORRECT ANSWER IS: (B)

7.

$$P = 50 \left(\frac{P}{G}, 1.5\%, 60 \right) = \$49{,}408$$

$P_{\text{After Taxes}} = 0.6 \, P = \$29{,}645$

THE CORRECT ANSWER IS: (C)

8.

Annual cost A = \$100,000 (A/P, 10%, 25) – \$25,000 (A/F, 10%, 25)

$\quad\quad\quad + \$15{,}000 + \$5{,}000$ (F/A, 10%, 10)(A/F, 10%, 25)

$\quad = 100{,}000 \, (0.1102) - 25{,}000 \, (0.0102) + 15{,}000 +$

$\quad\quad 5{,}000 \, (15.937)(0.0102)$

$\quad = 11{,}020 - 255 + 15{,}000 + 813 = \$26{,}578$

THE CORRECT ANSWER IS: (C)

9. $25,000 = 80,000 \, (A/P, \, \%, \, 20) - 20,000 \, (A/F, \, \%, \, 20) + 18,000$

Assume 10%

$7,000 \overset{?}{=} 80,000(0.1175) - 20,000(0.0175) = 9,050$

Assume 6%

$7,000 \overset{?}{=} 80,000(0.0872) - 20,000(0.0272) = 6,432$

Interpolate to find the annual rate of return (R):

$$\frac{R-6}{7,000 - 6,432} = \frac{10-6}{9,050 - 6,432}$$

$$\frac{R-6}{568} = \frac{4}{2,618}$$

$$R = \frac{(4)(568)}{2,618} + 6$$

$$= 0.86 + 6$$

$$= 6.86\%$$

THE CORRECT ANSWER IS: (A)

GENERAL AFTERNOON SOLUTIONS

10. Refer to the Engineering Economics section of the *FE Reference Handbook*.

At the end of 5 years, the car will be worth 5.8% of $21,000.

$21,000(0.058) = \$1,218$

THE CORRECT ANSWER IS: (B)

11. Refer to the Mechanics of Materials section of the *FE Reference Handbook*.

$$\delta = \frac{5w_o L^4}{384\,EI} = \frac{5(10,000)(5)^4}{384\left(200 \times 10^9\right)\left(21.58 \times 10^{-6}\right)} = 0.0189 = 18.9 \text{ mm}$$

THE CORRECT ANSWER IS: (D)

12. Refer to the Stresses in Beams portion of the Mechanics of Materials section of the *FE Reference Handbook.*

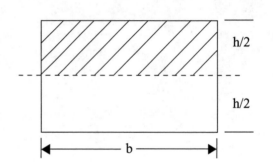

$$\tau_{xy} = \frac{VQ}{Ib} \qquad Q = A'y'$$

$$= \frac{h}{2}(b)\left(\frac{h}{4}\right)$$

$$= \frac{bh^2}{8}$$

$$\tau_{xy} = \frac{V\frac{bh^2}{8}}{\frac{bh^3}{12}b} \qquad = \left(\frac{3}{2}\right)\left(\frac{V}{bh}\right)$$

$$\tau_{xy} = \left(\frac{3}{2}\right)\frac{50 \times 10^3}{(12)(10)} = 625\frac{N}{cm^2} = 6.25 \text{ MPa}$$

$$\tau_{xy} = \left(\frac{3}{2}\right)\frac{100 \times 10^3}{(12)(10)} = 625\frac{N}{cm^2} = 6.25 \text{ MPa}$$

$$\tau_{xy} = \left(\frac{3}{2}\right)\frac{150 \times 10^3}{(12)(30)} = 625\frac{N}{cm^2} = 6.25 \text{ MPa}$$

THE CORRECT ANSWER IS: (B)

13. $\Sigma M_o = 0$

$P\delta - F\,(2\cos\theta) = 0$

$P\delta = F\,\sqrt{4 - \delta^2}$

$P = \dfrac{k\delta}{\delta}\,\sqrt{4 - \delta^2} = k\,\sqrt{4 - \delta^2}$

$P = 2k = 2(20) = 40\ kN$

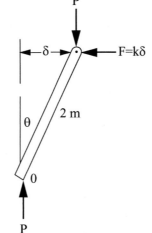

Check stability of AB:

$P = \dfrac{\pi^2 EI}{\ell^2}$

$P = \dfrac{\pi^2\left(200 \times 10^9\right)\left(20 \times 10^{-6}\right)}{2^2}$

$= 9.86\ MN$

Therefore $P_{max} = 40\ kN$

THE CORRECT ANSWER IS: (A)

14.   Carbon equivalent $= 0.25 + \dfrac{1.27 + 0.02}{6} + \dfrac{0.28 + 0.05 + 0.02}{10} = 0.50$

The graph indicates that with a carbon equivalent of 0.50 and a plate thickness of 37 mm both pre-heat and postheat are required.

THE CORRECT ANSWER IS: (C)

15. A36: carbon equivalent $= 0.25 + \dfrac{0.85 + 0.05}{6} + \dfrac{0.28 + 0.03 + 0.07}{10} = 0.44$

A517: carbon equivalent $= 0.09 + \dfrac{0.35 + 0.02}{6} + \dfrac{0.01 + 0.03 + 0.01}{10} = 0.16$

A572: carbon equivalent $= 0.23 + \dfrac{1.65 + 0.01}{6} + \dfrac{0.30 + 0.08 + 0.03}{10} = 0.55$

A715: carbon equivalent $= 0.05 + \dfrac{1.65 + 0.80}{6} + \dfrac{0.90 + 0.07 + 0.06}{10} = 0.56$

THE CORRECT ANSWER IS: (B)

16. Convert from wt% to vol%

70 g Cu + 30 g Cr

$$\text{vol\% Cu} = \dfrac{\dfrac{70\,g}{8.93\,g/cm^3}}{\left(\dfrac{70\,g}{8.93\,g/cm^3} + \dfrac{30\,g}{7.19\,g/cm^3} \right)} \times 100 = 65.26$$

vol% Cr = 34.74

Calculate the upper limit of ρ_c

$$\rho_U = \left[(65.26)(1.70) + (34.74)(12.9) \right]/100 = 5.59 \; \mu\Omega\text{·cm}$$

Calculate the lower limit of ρ_c

$$(\rho_L)^{-1} = \left[\dfrac{65.26}{1.70} + \dfrac{34.74}{12.9} \right]/100$$

$\rho_L = 2.43 \; \mu\Omega\text{·cm}$

$2.43 \; \mu\Omega\text{·cm} \leq \rho_c \leq 5.59 \; \mu\Omega\text{·cm}$

THE CORRECT ANSWER IS: (B)

17. $V_1 = 60 \text{ m/s}$

$T_2 = T_1$

$P_1 = 600 \text{ kPa}$ $\qquad\qquad$ $P_2 = 300 \text{ kPa}$

$\dot{m}_2 = \dot{m}_1$

$$\frac{P_2 \dot{V}_2}{RT_2} = \frac{P_1 \dot{V}_1}{RT_1} \qquad \text{or} \qquad P_2 A V_2 = P_1 A V_1$$

$$V_2 = V_1 \frac{P_1}{P_2} = 60 \left(\frac{600}{300}\right) = 120 \text{ m/s}$$

THE CORRECT ANSWER IS: (C)

18. Volume flow rate = area × velocity

$Q = Av$

$$v = \frac{Q}{A}$$

$$\left(\frac{40 \text{ m}^3}{\text{min}}\right)\left(\frac{\text{min}}{60 \text{ s}}\right)\left(\frac{1}{\pi (0.15)^2 \text{ m}^2}\right) = 9.4 \text{ m/s}$$

THE CORRECT ANSWER IS: (A)

19. From the Moody (Stanton) diagram in the Fluid Mechanics section of the *FE Reference Handbook* with $\dfrac{e}{D} = 0.0005$ and $\text{Re} = 2 \times 10^5$

Then, f is 0.019

THE CORRECT ANSWER IS: (B)

20. $P = \gamma h$

$$= \left(9,800 \text{ N/m}^3\right)(9 \text{ m}) = 88,200 \text{ N/m}^2$$

$$= 88,200 \frac{\text{N}}{\text{m}_2} \frac{\text{Pa} \cdot \text{m}^2}{\text{N}} = 88,200 \text{ Pa}$$

$$= 88.2 \text{ kPa}$$

THE CORRECT ANSWER IS: (C)

21.

$$V = Q/A = \frac{0.5 \dfrac{m^3}{min} \dfrac{min}{60\,s}}{\pi \dfrac{(0.075\ m)^2}{4}} = 1.89\ m/s$$

$$h_L = f \frac{L}{D} \frac{v^2}{2g} = 0.02 \frac{100\ m}{0.075\ m} \frac{(1.89\ m/s)^2}{2\,(9.81\ m/s^2)}$$

$$= 4.8\ m$$

$$P = \frac{\gamma Q h}{e}$$

$$= \frac{\left(1{,}000\,\dfrac{kg}{m^3}\right)\left(9.81\,\dfrac{m}{s^2}\right)\left(0.5\,\dfrac{m^3}{min}\right)\left(\dfrac{min}{60\,s}\right)(10.5 + 4.8\ m)}{0.8}$$

$$= 1{,}563 \frac{kg \cdot m^2}{s^3}$$

$$= 1{,}563\ W$$

THE CORRECT ANSWER IS: (C)

22. Inductor current will not change instantaneously at the instant the switch closes.

$$I = \frac{1}{2}10\,mA = 5\,mA$$

THE CORRECT ANSWER IS: (B)

23. Node 1: $\dfrac{V_1 - 100}{10} + \dfrac{V_1}{20} + \dfrac{V_1 - V_2}{30} = 0$

$11V_1 - 2V_2 = 600$

Node 2: $\dfrac{V_2 - V_1}{30} + \dfrac{V_2}{40} = 0$

$-4V_1 + 7V_2 = 0$

Solving, $V_1 = \dfrac{7}{4}V_2$

$11\left(\dfrac{7}{4}V_2\right) - 2V_2 = 600 \Rightarrow V_2 = 34.78 \text{ V}$

THE CORRECT ANSWER IS: (B)

24. Node 1: $\dfrac{V_1 - 10}{20} + \dfrac{V_1 - V_2}{20} + \dfrac{V_1}{20} = 0$

Node 2: $\dfrac{V_2 - V_1}{20} + \dfrac{V_2 - 10}{20} + \dfrac{V_2}{20} = 0$

$3V_1 - V_2 = 10$

$-V_1 + 3V_2 = 10$

Solving, $V_1 = 3V_2 - 10$

$8V_2 = 40$

$V_2 = 5 \text{ V}$

$V_1 = 5 \text{ V}$

THE CORRECT ANSWER IS: (C)

25. When fouling occurs in a heat exchanger, the overall heat-transfer coefficient is given by the following equation from the Heat Transfer section of the *FE Reference Handbook*.

$$\frac{1}{U_o\,A_o} = \frac{1}{h_i\,A_i} + \frac{R_{fi}}{A_i} + \frac{t}{k\,A_{avg}} + \frac{R_{fo}}{A_o} + \frac{1}{h_o\,A_o}$$

The two fouling terms cannot be determined separately, and they will be combined into the following:

$$\frac{R_f}{A_o} = \frac{R_{fi}}{A_i} + \frac{R_{fo}}{A_o}$$

When the tube-wall resistance is neglected, R_f can be calculated.

$$R_f = \frac{1}{U_o} - \frac{D_o}{h_i\,D_i} - \frac{1}{h_o}$$

$$= \frac{1}{700\ \text{W}/\left(\text{m}^2 \cdot \text{K}\right)} - \frac{25\ \text{mm}}{\left[2{,}500\ \text{W}/\left(\text{m}^2 \cdot \text{K}\right)\right](19\ \text{mm})} - \frac{1}{1{,}500\ \text{W}/\left(\text{m}^2 \cdot \text{K}\right)}$$

$$= 0.00024\ \text{m}^2 \cdot \text{K/W}$$

THE CORRECT ANSWER IS: (B)

26. The steady-flow energy equation for the turbine, neglecting kinetic and potential energy changes, is:

$q - w = h_2 - h_1,\ q = 0$

$w = h_1 - h_2 = 2{,}960.70 - 2{,}237.40 = 723.3\ \text{kJ/kg}$

THE CORRECT ANSWER IS: (B)

27. An energy balance on the condenser gives:

$$q - w = m_5 h_3 + m_{cw} h_{out} - m_5 h_2 - m_{cw} h_{in}$$

since $q = w = 0$

$$m_{cw} (h_{out} - h_{in}) = m_5 (h_2 - h_3)$$

$$\frac{m_{cw}}{m_5} = \frac{h_2 - h_3}{c(T_{out} - T_{in})} = \frac{2,237.40 - 251.13}{4.18(32 - 21)} = 43.2$$

THE CORRECT ANSWER IS: (C)

28. For the boiler, the steady-flow energy equation gives:

$$q - w = h_5 - h_4, \; w = 0$$

$$q = 2,715.70 - 255.18 = 2,460.5 \text{ kJ/kg} = 2.46 \text{ MJ/kg}$$

THE CORRECT ANSWER IS: (A)

29.   Isothermal process: PV = mRT

$$P_2V_2 = P_3V_3$$

$$V_3 = \frac{P_2V_2}{P_3}$$

$$_3Q_2 = \int_2^3 \delta Q = \int_2^3 PdV = mRT_1 \ln\frac{V_3}{V_2} = P_2V_2 \ln\frac{V_3}{V_2}$$

$$\ln\frac{V_3}{V_2} = \frac{(60)(1,000)}{P_2V_2}$$

$$= \frac{60,000}{(9\times10^5)(0.3)} = \frac{0.6}{2.7} = 0.2222$$

$$\frac{V_3}{V_2} = e^{0.2222} = 1.2488$$

$$V_3 = (0.3)(1.2488) = 0.375\,\text{m}^3$$

THE CORRECT ANSWER IS: (A)

30. $Q_H = 60$ kJ

$\eta_{th} = 0.48$

$W_{out} = (60)(0.48) = 28.8$ kJ

$Q_L = Q_H - W = 60{-}28{=}31.2$

THE CORRECT ANSWER IS: (C)

FE Study Material Available for Purchase

FE Supplied-Reference Handbook

Sample Questions and Solutions (in printed and a multidiscipline, CD-ROM format)
are available for the following modules:

Chemical
Civil
Electrical
Environmental
General
Industrial
Mechanical

**For more information about these and other Council publications and services,
visit us at www.ncees.org or contact our
Customer Service Department at (800) 250-3196.**